SOTHIS PRESS

Sothis Press aims to completely reimagine the public domain book with an emphasis on universal access to captivatingly designed print editions. We believe that technology will enable every single book ever published to be perpetually accessible in beautiful print form. Sothis is beauty in ubiquity.

www.sothis.press
1 Waterhouse Square, London, ECIN 2ST

The Analysis and Design of Indicator Reducing Motions

Thesis for the Degree of Bachelor of Science in Mechanical Engineering; College of Engineering, University of Illinois, 1912

Samuel Elza Gregg

SOTHIS PRESS

ISBN 978-1-0365-0973-6

The Analysis and Design of Indicator Reducing Motions is in the public domain.

This edition is published by Sothis Press © 2023. The company is registered in England and Wales. The Company Number is 14900389.

THE ANALYSIS AND DESIGN OF

INDICATOR REDUCING MOTIONS

BY

SAMUEL ELZA GREGG

THESIS

FOR THE

DEGREE OF BACHELOR OF SCIENCE

IN

MECHANICAL ENGINEERING

COLLEGE OF ENGINEERING

UNIVERSITY OF ILLINOIS

1912

UNIVERSITY OF ILLINOIS

May 31st, 1912. 190

THIS IS TO CERTIFY THAT THE THESIS PREPARED UNDER MY SUPERVISION BY

Samuel Elza Gregg

ENTITLED The Analysis and Design of Indicator Reducing Motions

IS APPROVED BY ME AS FULFILLING THIS PART OF THE REQUIREMENTS FOR THE

DEGREE OF Bacherlor of Science in

Mechanical Engineering.

W. V. Dunkin

Instructor in Charge

APPROVED: C. R. Richards

HEAD OF DEPARTMENT OF Mechanical Engineering.

TABLE OF CONTENTS.

I INTRODUCTION

The steam engine indicator furnishes us a means of determining the amount of work done in the cylinder of an engine by the steam. There are many forms of indicators, two of which, the Crosby and the Thompson, are shown on the following page. The principle upon which each works, however, is the same in all cases.

When the indicator is connected to the cylinder of the engine; the steam comes in contact with the piston in the small cylinder of the indicator. A coil spring resists the movement of the indicator piston and as the spring is compressed, motion is transmitted through a series of links, to a pencil point. The rotation of the indicator drum is proportional to the stroke of the engine and the pencil point traces a curve on a card wrapped around the drum. The ordinates of this curve represent to some scale the pressure of the steam in the engine cylinder. By calibrating this spring the scale may be easily determined.

The most important requirement for the proper operation of the indicator is, that the rotation of the drum be proportional to the stroke of the engine at every point of the stroke. This rotation is obtained by means of a cord, one end of which is wrapped around the base of the drum. The other end of the cord is attached, through a reducing motion, to the cross-head of the engine. If one end of the cord should be fastened directly to the cross-head of the engine the circumference of the drum and consequently the length of the card would be equal to the stroke of the engine. This length of

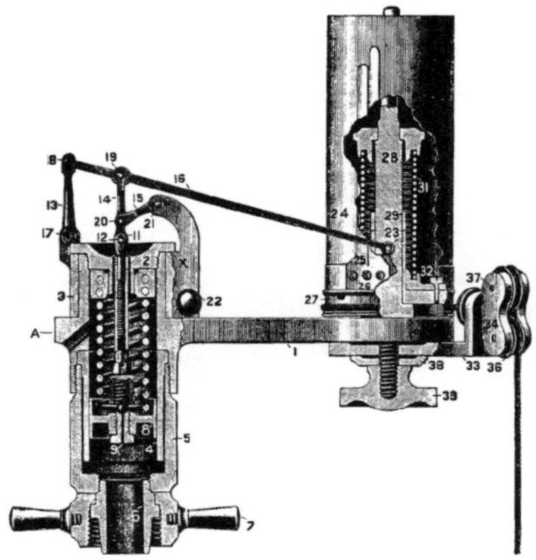

Crosby Indicator

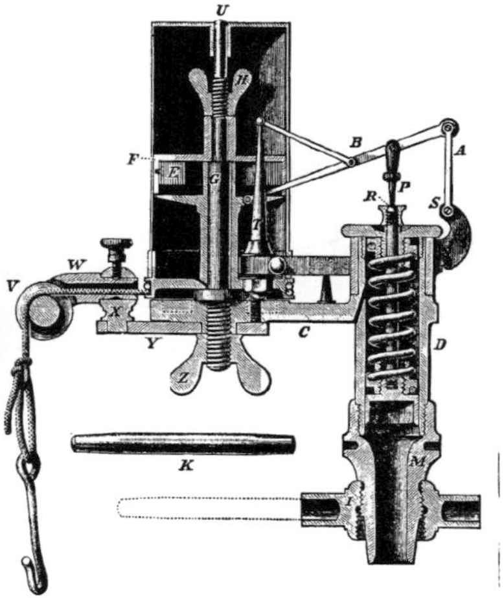

Thompson Indicator

card is neither practical nor desirable and engineers were quick to conceive of the idea of reproducing the motion of the cross-head to a smaller scale. The mechanism used for this purpose is called the indicator rig or reducing motion.

Experience and general practice now calls for a card from three to five inches long. With this length of card it is important that the motion be reproduced exactly.

Reducing motions of almost every conceivable form have been designed to meet the needs of the testing engineer and to conform to the different types of engines now in use. Some of these motions are mathematicaly correct, that is, every point of the reduced motion corresponds exactly to a similar point on the original motion, while in others this is only approximately true. The former are known as exact and the latter as approximate motions.

The indicator card is used for determining a large number of facts about the steam engine, such as point of cut off, release, compression, admission, the weight of steam in the cylinder at any point of the stroke and the indicated horse power. Now, if the reducing motion does not reproduce these motions to correspond exactly to the points on the stroke, it is seen that an error will result in determining these things by means of the indicator card.

As the ratio of the reduced to the original motion must necessarily be large for long stroke engines the importance of exact reduction becames evident. In gas engine work the indicator card is not used to determine the events of the stroke and small errors in the reducing motion are not so serious as in the case of the steam engine. Therefore, it is possible to use one of the approximate motions for gas engine work, provided the error is not too great. The limit of the allowable error has never been determined but in all cases it should be made as small as possible.

In the following pages a few of the different reducing rigs are illustrated, analyzed and discussed with reference to their advantages and disadvantages. Several designs are shown and an attempt is made to bring out some of the more important points which have to be considered in the design of any logical form of reducing motion.

II ANALYSIS OF THE BRUMBO-PULLEY

The Brumbo-pulley is used extensively in gas engine work because of the simplicity and ease of attachment to the engine. It gives an approximate motion only, but modern gas engines are of such construction that it is difficult to attach any of the other forms of reducing motions. It consists of three parts or links, the oscillating pendulum OA in Fig. 1 is fixed at O and is connected to the moving piston by the link AB, a sector C is fastened to the link OA near the fixed end and from this the cord leads to the indicator drum. Raising or lowering the fixed point will diminish or increase the length of the card as desired.

The mechanism shown in Fig. 1 is drawn to $\frac{1}{8}$ scale, the dimensions being taken from a rig now in use on a Meyer automatic engine of 8" bore and 12" stroke. The length of the link AB is 7", of OA 17" and of DO 4", O being the fixed point and B the point on the cross-head. The rig is placed on the engine so that the link AB is horizontal, or parallel to the stroke, when the cross-head is in mid-position.

In Fig. 2 lay off OA, OD and AB according to the above dimensions. If B represents the middle of the stroke and we make BB_1 and BB_8 each equal to 6" B_1B_8 will then be equal to 12", the length of the stroke. When the engine is on the head end dead center, the mechanism will be in the position OA_1B_1, and in the position OA_8A when the cross-head moves to the opposite of the stroke. As the cross-head moves from B_1 to B_8 the arm OA swings

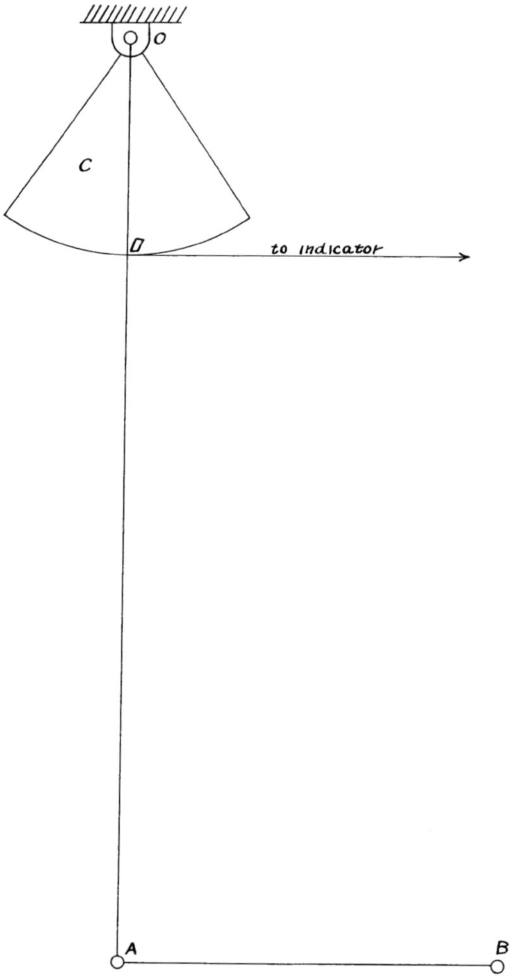

Fig. 1.

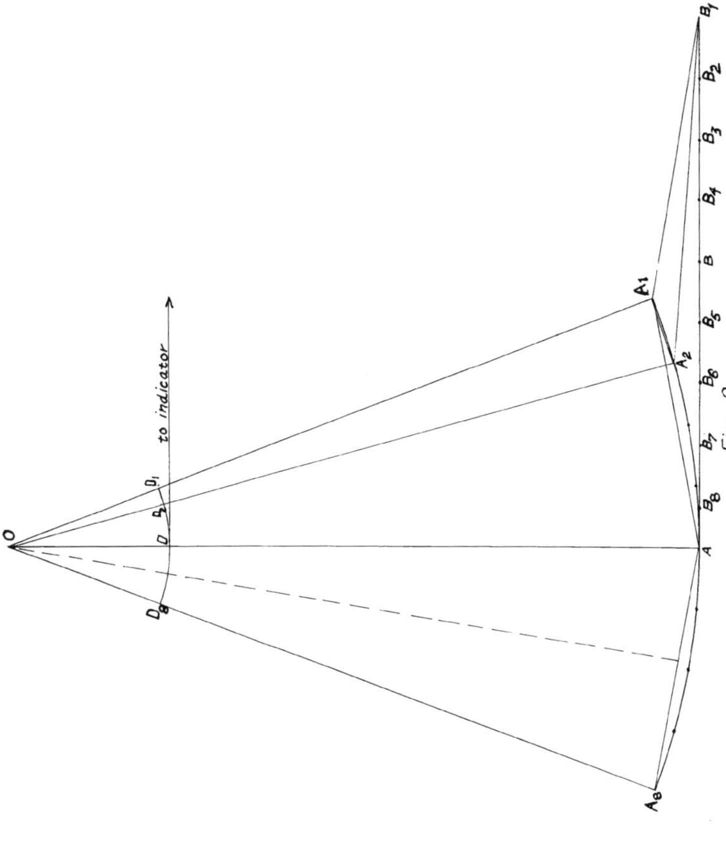

Fig. 2.

through the angle A_8OA_1, the point D will have moved from D_1 to D_8 and the length of the arc D_1DD_8 will be equal to the length of the cord unwound off the indicator drum, neglecting the stretch which might occur in the cord itself. The arc D_1DD_8 is also equal to the length of card which would be obtained by the use of this rig. The length of the arc D_1DD_8 will be equal to the radius OD times the angle A_1OA_8 when this angle is expressed in radians. The most accurate method of determining this angle is to connect the point A with A_8 and A1 by the lines AA_8 and AA_1, scale off the length of these lines and the sin of $\frac{1}{2}$ the angle A_8OA will be equal to $\frac{1}{2}A_8A/17$. In a similar manner sin $\frac{1}{2}$ the angle $A_1OA = \frac{1}{2}A_1A/17$. The value of these angles may be looked up in any table of trigonometric functions and their sum will be the required angle A_1OA_8. Dividing this sum by 57.296 to reduce the angle to radians and multiplying by OD we have the length of the arc D_1DD_8 which is the length of the indicator card.

For convenience let the length of the stroke be divided into 8 equal parts and the cross-head be assumed to be in the position B_1. If we now let the cross-head move over 1/8 of the stroke the reducing motion will be in the position OA_2B_2. The ratio of the cross-head travel to the length of the stroke is $B_1B_2/B_1B_8 = 1/8$. The ratio of the travel of a point on the indicator drum to the length of card is A_1A_2/A_1A_8. For an exact reproduction of the cross-head motion these ratios should be equal to each other and if they are not there is an error in the reducing motion equal to the difference in the two ratios. The ratio A_1A_2/A_1A_8 is equal to the ratio of the angle A_1OA_2 to the angle A_1OA_8. A method of determining the angle GOH has already been outlined and a similar one may be used for measuring the angle A_1OA_2. Connect the points A_1 and A_2 by the line A_1A_2, then the sin of ½ the angle A_1OA_2 = $\frac{1}{2}A_1A_2/A_1O$. These distances may be scaled off the diagram and the values of the angles found by the use of the trignometric tables as before.

In this particular case OA_8 = OA_1 = 17" and AA_8 = 6". Then the sin ½ A_8OA = ½ x 6/17 = 0.1765 from which the angle A_8OA = 20.35° AA_1 = 6.91". sin ½ AOA_1 = ½ x 6.91/17 = 0.1821 from which the angle AOA_1 = 21°. Therefore, the angle A_8OA_1 = 21. + 21.35 = 41.35° 41.35/57.296 = 0.7225, 0.7225 x 4 = 2.89" which is the length of the card that would be obtained. A_1A_2 = 1.64", then sin $\frac{1}{2}A_1OA_2$ = ½ x 1.64/17 = 0.04825 from which the angle A_1OA_2 = 5.533°. Therefore the ratio of the arc D_1D_2 to the arc D_1D_8 is equal to 5.533/41.35 = 0.134 which is the ratio of the travel of the point on the indicator drum to the length of the card. By making a similar calculation for each of the eight points on the stroke the following table may be constructed.

TABLE 1.

Point	Ratio C.H. travel to stroke.	Ratio drum travel to card length.	Difference of ratios.	Error in % length of card.
B_1	0.00	0.00	0.00	0.00
B_2	0.125	0.134	0.009	0.9
B_3	0.25	0.2615	0.0115	1.15
B_4	0.375	0.384	0.009	0.9
B	0.5	0.507	0.007	0.7
B_5	0.625	0.629	0.004	0.4
B_6	0.75	0.752	0.002	0.2
B_7	0.875	0.876	0.001	0.1
B_8	1.00	1.00	0.00	0.00

From the above table it can be seen that there is a large

error in some of the ratios and that they are equal only at the end positions. The greatest error is found to be at one quarter stroke where it amounts to 1.15% of the length of the card. The errors need not be calculated when the piston is going in the opposite direction since they have the same numerical values and may be read from the bottom of the table upwards. These errors are excessive and a card taken with any such reducing motion is practically useless for either steam or gas engine work.

A slight change in the design of the mechanism will reduce these errors to a minimum. Increase the length of the arm OA so that the point A travels equal distances above and below the horizontal line B_1B_8 and locate the point O, by trial, so that the angles A_8OA and A_1OA will be equal. Fig. 3 shows the mechanism after these corrections have been made. The errors may now be calculated as before and the angle A_8OA will be 39.93 4° from which the length of the card is found to be 2.784". This length may be increased if desired by increasing the radius of the sector D_8OD_1. After making these changes the errors are as shown in the following table.

TABLE 2

Point.	Ratio C.H. travel to stroke.	Ratio drum travel to card length.	Difference of ratios.	Error in % length of card.
B_1	0.00	0.00	0.00	0.00
B_2	0.125	0.131	0.006	0.6
B_3	0.25	0.2561	0.0061	0.61
B_4	0.375	0.3915	0.0065	0.65
B	0.5	0.5	0.00	0.00
B_5	0.625	0.628	0.003	0.3
B_6	0.75	0.751	0.001	0.1
B_7	0.875	0.8755	0.0005	0.05
B_8	1.00	1.00	0.00	0.00

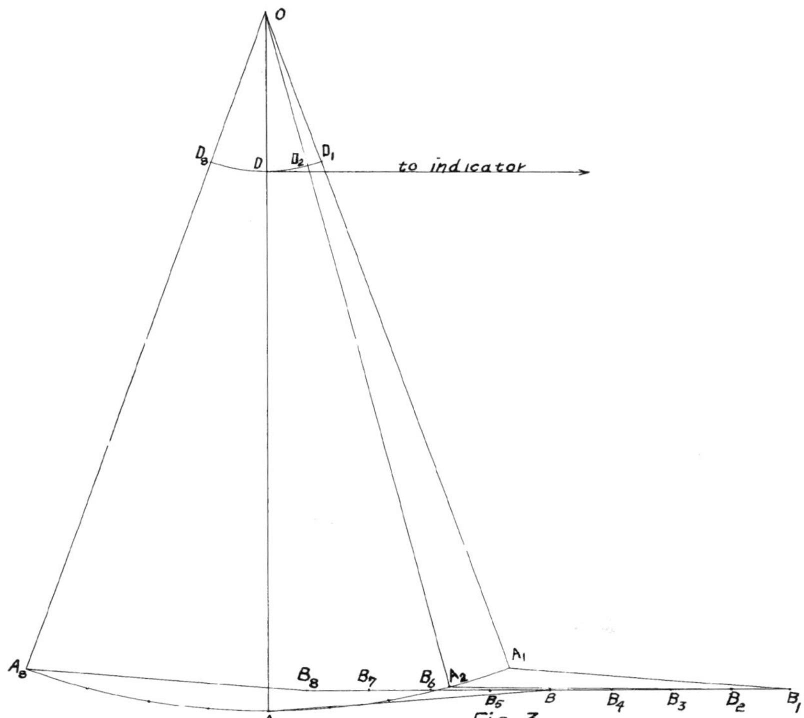

Fig. 3.

The greatest error as shown by table 2 is found to be at $\frac{3}{8}$ stroke where it amounts to 0.65% of the length of the card. This error is one which would sometimes cause trouble in steam engine.

The mechanism may be made to produce a more accurate motion by increasing the length of the arm AB since the angularity of this arm decreases rapidly with the increased length.

The results tabulated in tables 1 and 2 are shown more clearly by the following curves which are plotted to the same scale so that the errors may be compared at a glance.

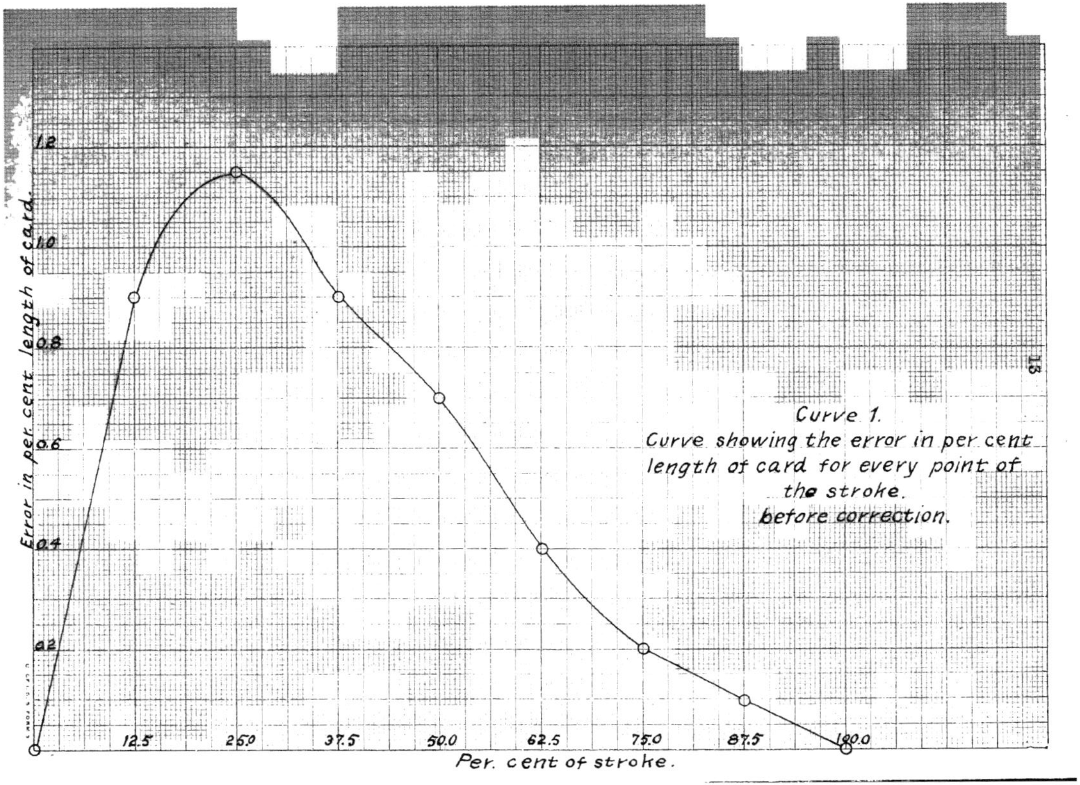

Curve 1.
Curve showing the error in per cent length of card for every point of the stroke.
before correction.

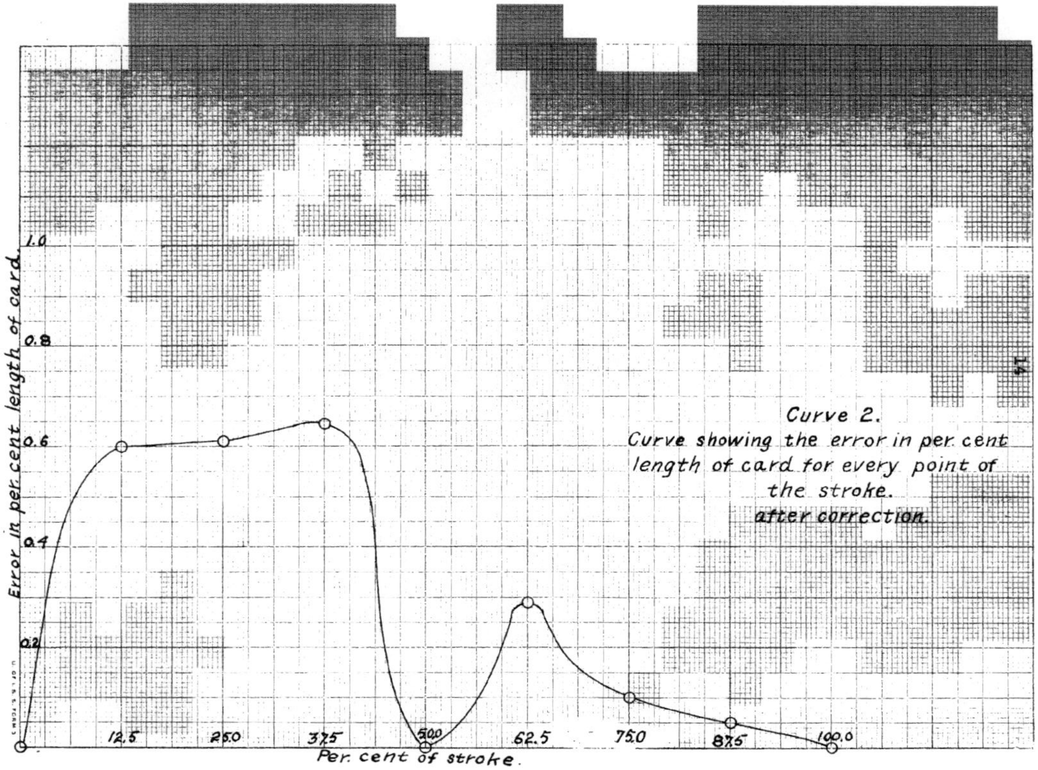

Curve 2.
Curve showing the error in per. cent length of card for every point of the stroke.
after correction.

III ANALYSIS OF THE INCLINED PLANE-BELL CRANK MOTION.

Another form of reducing motion which has been used is that shown in Fig. 4. This particular one was designed by Mr. J. W. Grant of the Graton and Knight Manufacturing Co. of Worchester Mass. It consisted of an inclined plane bolted to the cross-head as shown and a bell-crank pivoted at B. The arm BE is held in contact with the plane by means of a coil spring not shown in the figure. The motion is a very simple one, easy to construct and attach to the engine, but there are some few points which must not be overlooked if the reduced motion is to be of the required accuracy.

When the pivot B is placed above the point of contact E, as in Fig. 4, a lever motion is obtained which does not reproduce the motion of the piston correctly. In the figure the point of contact E of the arm BE represents the middle point of the stroke; the point would drop to C at the end of the stroke and rise to the point D at the other end. The angular movement of the lever arm E to C is seen to be much greater than that from E to D. This movement being reproduced at the top of the lever which communicates the motion to the indicator drum, a diagram is produced which is shortened at one end and lengthened at the other. The shortened end is represented by ed and the lengthened end by ec.

When this sliding contact is used the point B, should be dropped to the point F on the same level with the point of contact E of the arm BE when the cross-head is in the middle of its stroke. If a roller contact is to be used the point F should of course, be

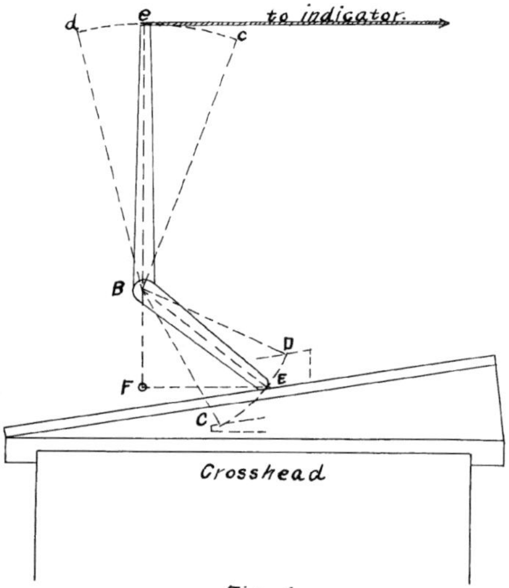

Fig. 4.

on a level with the center of the roller. The American Engine Co., of Bound Brook N. J. use this kind of a reducing motion on their Ball engines with very good success. If the arm BE is short an off-set might be required to prevent the pivot point F from striking the plane when the arm BE rises to its highest position BD. If the vertical movement of the point E does not exceed ½ of the length of the arm FE the motion imparted to the indicator drum will be sufficiently accurate for all ordinary purposes. If either the arm FE or the string to the indicator is short a sector should be provided, as in the case of the Brumbo-pulley, to correct for the angularity of these parts. In case it is necessary to keep the pivot point B above the level of the point of contact E an arrangement such as shown in Fig. 5, which was suggested by Mr. Charles W. Barnaby, may be used. In this case the vertical rod C, carrying the wheel E at its lower end, works in the guide D, which is fastened to the engine frame. A sector F is provided at the end of the arm BF and the vertical motion of the rod C is then transmitted through the bell-crank FBe to the indicator drum. The vertical motion of C is proportional to the travel of the cross-head and if a sector similar to F is provided at e the reduced motion will be an exact reproduction of the motion of the piston, neglecting any lost motion or stretch of the string which might occur. For high speed engines an auxilliary spring has to be used to overcome the inertia of the parts and to keep the wheel E in contact with the plane, but in the case of slow speed engines the tension of the spring in the indicator drum itself is sufficient for this purpose.

The arm AB is provided so that the motion may be held stationary when the rig is not in use. A hole should also be

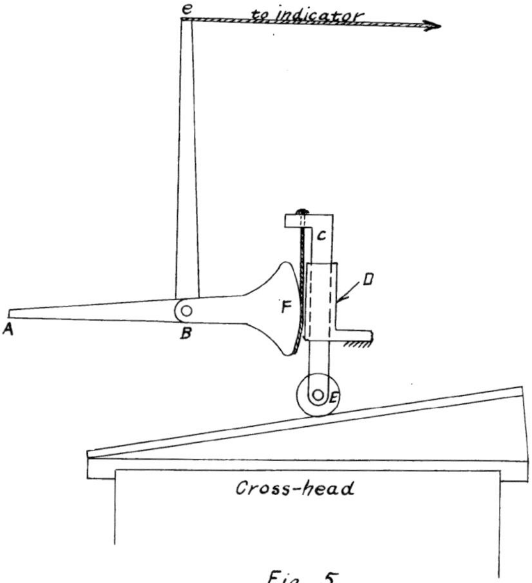

Fig. 5.

drilled in the slide D into which a pin could be dropped to hold the wheel E up off the plane when the mechanism is not in use.

Another way in which this device can be made to give the correct motion is to put in place of the inclined plane a curved surface which will give the exact motion to the bell-crank. The method of laying out such a curve is illustrated in Fig. 6. Instead of having the slide move back and forth under the lever, the lever will be considered as moving back and forth above the slide. We give to the lever the intended perfect motion, and it traces on the plane of the slide the profile of the curve which will produce this motion. In order to do this the method suggested by Prof. R. C. H. Heck will be used, where MP, Fig. 6, is the working lever which is to oscillate through the equal angles MPa and MPb from mid-position. Divide the path of P, from a to b, into any number of equal parts, say eight. Draw through M a line AB, parallel and equal to the stroke of the cross-head, with M as its mid-point, and divide this into the same number of equal parts, (eight), so that the divisions on AB and ab will correspond. Take an angular position of the lever as Mb, and carry it out to its proper location on the stroke line, at B, where it appears as BE, parallel to MB.

This is done most easily by drawing a horizontal line bE, and making the length bE equal to MB. The intersection of BE and bE locates the center of the small roller when the mechanism is in that position. With E as a center draw a small circle to represent the roller on the end of the arm MP. The roller may be omitted and a round point used if desired. The same process is carried out for each of the remaining points and by drawing a curve tangent to the

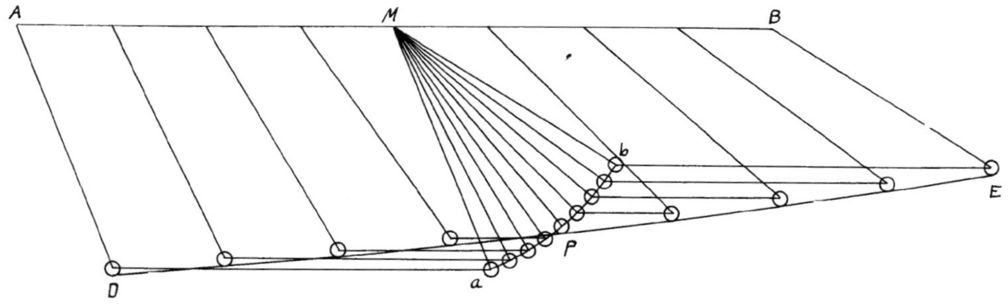

Fig. 6.

small circles we will have the shape of the slide required for giving an exact reproduction of the motion of the cross-head. There will still be a slight error unless a sector is provided at the opposite end of the bell-crank to correct for the angularity of the cord. Only one arm of the bell-crank is shown in the Fig. but if one end is given the required motion, the other end will receive the same motion since the link is a rigid one.

The objection to this scheme is the difficulty of making the special curved slide.

IV ANALYSIS OF THE CURVED BELL-CRANK.

Instead of placing an inclined plane on the cross-head of the engine one of the arms of the bell-crank may be given such a curvature that the motion of the piston may be reproduced exactly, if the curved arm of the bell-crank is held in contact with a pin, or small roller, attached to the cross-head of the engine. The manner of laying out a curve which will give the exact motion is shown in Fig. 7.

Let dg represent the length of the stroke and the arc AB the length of the card desired. Let O be the fixed point, which may be located at any convenient point on the engine frame, and OC equal the length of the working lever. Divide the stroke into any number of equal parts, say ten, and draw a line perpendicular to the stroke through each of these division points. With O as a center and OC' as a radius strike off the arc ab, making the angle bOa equal to the angle BOA, and divide it into the same number of equal parts (ten). The intersection of a line perpendicular to the stroke, drawn through any point on the stroke, with a line passing through O and a corresponding point on the arc ab will locate one point on the curve. Each of the other points may be located in a similar manner and a small circle drawn at each point to represent the roller on the cross-head. A curve drawn tangent to all of the positions of the roller gives the shape of the desired slide.

The curves OC'A' and OC'A" are symetrical about a horizontal axis and either may be used to produce the desired motion. A

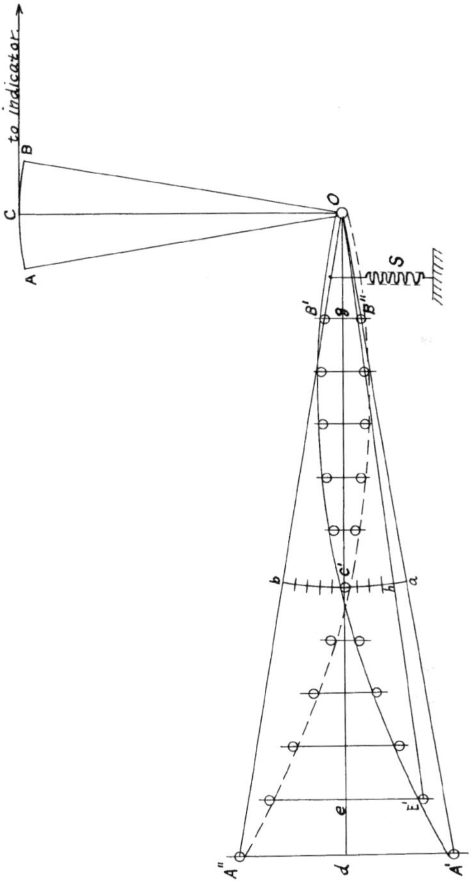

Fig. 7.

spring S is provided to keep the curved surface in contact with the roller C' when the spring in the indicator drum itself is not sufficient for this purpose or when the cord pulls the working lever in the opposite direction. A sector should be provided at the end of the working lever to correct for the angularity of the cord.

The disadvantages of this form of reducing motion are; the difficulty of constructing the special curved bar; each set of proportions requires a particular curve; it is not suitable for high speeds on account of inertia, and for these reasons it is not desirable form of reducing motion to use. The effect of inertia may be overcome by providing a slot to guide the pin C', however this would make the rig more difficult to construct and would interfere with holding the slide out of contant with the roller when the mechanism was not in use.

V THE DESIGN OF A PANTAGRAPH.

It is required to design a reducing motion for the Chandler and Taylor engine in the Mechanical Engineering Laboratory at the University of Illinois. The engine is of the double cylinder, horizontal type having 11" x 14" cylinders and is at present used to operate a brake shoe testing machine. The most convenient place to tap for indicator cocks is on the side of the cylinder. The design of the engine frame is such that ample room is available for the attachment of the rig in line with the indicator cock. Forged connections will be used for fastening the rig to the upper guide and to the cross-head of the engine.

The simplest form of a pantagraph is the parallelogram, or the four link chain, with opposite links equal. Such a mechanism is shown in Fig. 8. To prove that it will give the required straight line motion let O be the fixed point, draw any line through O, as OCA, cutting the non-adjacent links in C and A. Then in whatever position the mechanism be placed, these three points will always lie in a straight line. For by similarity of the triangles GOA and FCA

$$FC : GO :: FA : GA$$

$$FC = \frac{FA \times GO}{GA}$$

and therefore C is a fixed point on the link EF. The ratio OA : OC equals GA : GF and is therefore constant for all positions of the mechanism. From this it follows that if the point A traces any curve the point C will describe a similar curve on a reduced scale.

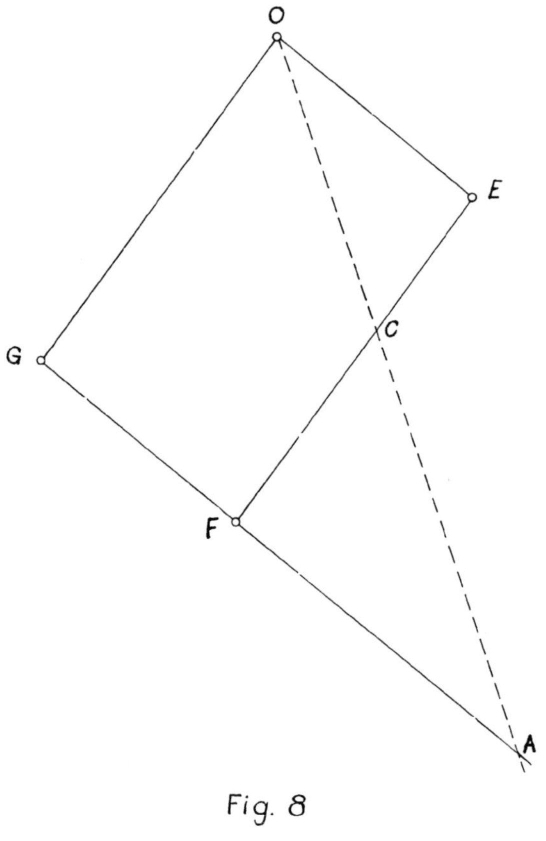

Fig. 8

This ratio can be either increased or diminished by moving C toward F or toward E.

In order to design a reducing motion of this type lay off AB in Fig. 9 equal to the length of the stroke, 14", to a convenient scale. Also lay off CD equal to the desired length of the card. The length may vary from 3" to 5" and for this engine $3\frac{1}{8}$" will be chosen. The location of the line CD with respect to the line AB is arbitraily chosen. Draw straight lines throough AC and BD and locate the point O at their intersection. From A let any line as AG be drawn and from O any other line as OG intersecting the line AG at the point G. Then through C draw a line parallel to OG intersecting AG at F. Complete the parallelogram by drawing OE. Now according to the above theory if A is fastened to the cross-head of the engine and starts to move toward B, the point C will move along the line CD and will reach D when A coincides with B which is the opposite end of the stroke. The point O must be fixed to the frame of the engine in such a position that the point C will lie in the same horizontal plane as the cord from the indicator drum.

Fig. 10 shows the pantagraph ready to attach to the engine, Fig. 11 contains the details of each part, while in Fig. 12 is shown the engine connections. The following is the bill of material necessary for its construction.

BILL OF MATERIAL.

1 soft steel bar, 1" x 1/8" x 3' - 8"

1 soft steel bar, 1" x $\frac{1}{4}$" x 2' - 5"

1 hard steel rod, $\frac{7}{8}$" x 6"

11 brass washers, outside diameter, 1", inside diameter $\frac{1}{2}$", thickness 1/32".

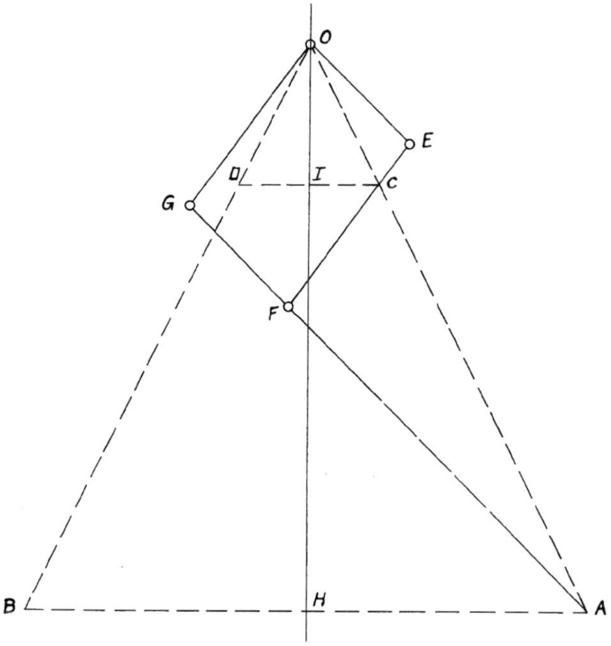

Fig. 9

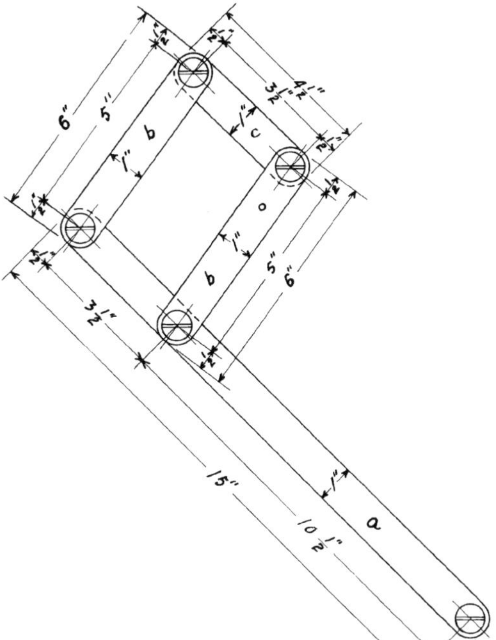

Fig. 10.
Pantagraph Reducing Motion.

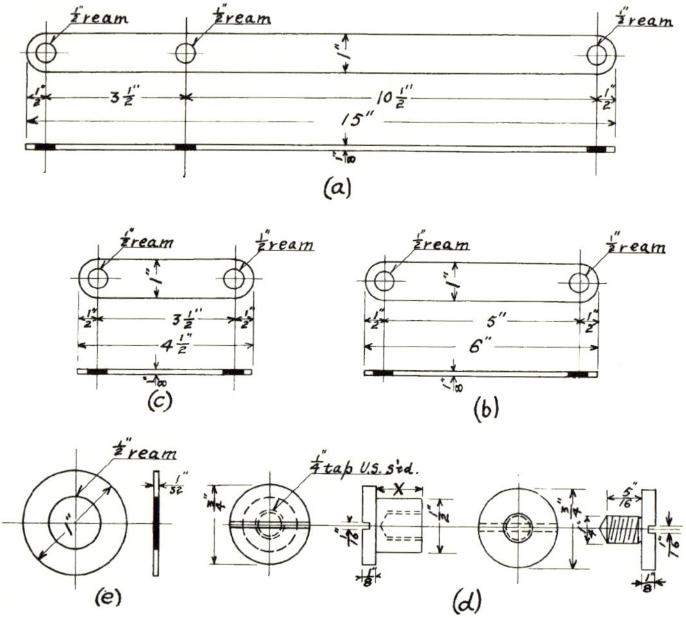

Note:

(a) Soft Steel, Make 1.

(b) Soft Steel, Make 4.

(c) Soft Steel, Make 1.

(d) Hard Steel, Make 4, dimension X = 7/16"

(d) Hard Steel, Make 1, dimension X = 19/32"

(e) Brass, Make 11.

Fig. 11.

Details of Pantagraph Reducing Motion.

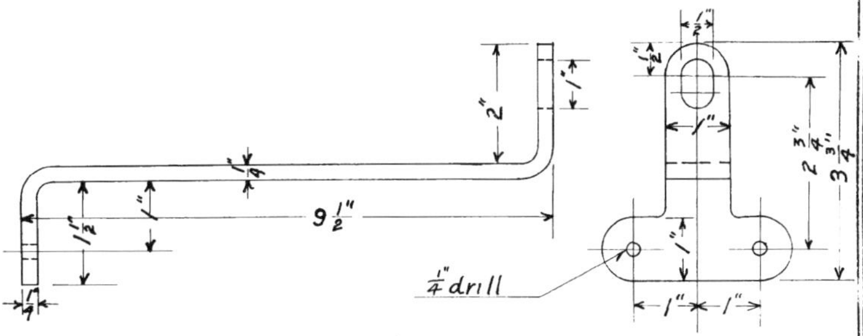

Upper Guide Connection.

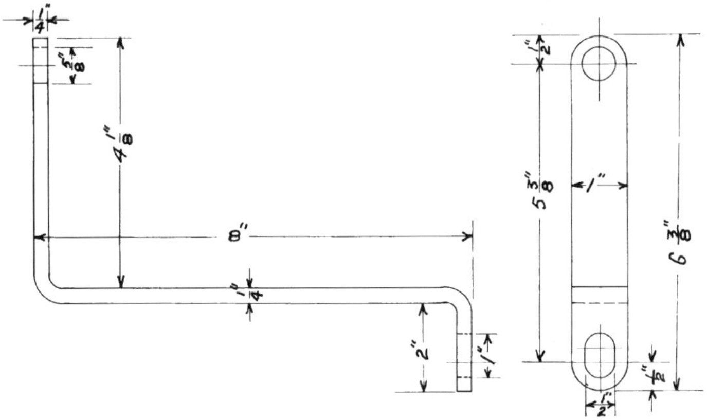

Crosshead Connection.

Fig. 12.

Brackets for attaching Pantagraph Reducing Motion to the Engine.

VI THE DESIGN OF A REDUCING MOTION FOR A CORLISS ENGINE

The indicator rig shown in Fig. 13 is similar to the Brumbo-pulley but has the extra link pc which corrects for the angularity of the link PC and makes the reduced motion an exact reproduction of the cross-head motion. The following design is intended to be used on a 12" x 24" Allis-Corliss engine. If we desire a card 4" long a 6 to 1 reduction will be necessary.

Let O be the fixed point located in any convenient position, C a point on the cross-head and AB the length of the stroke. For ease in construction the link pc should not be less than 3" in length, hence for the given reduction OP must equal 6 times Op.

Now if we locate p at a point 1/6 of the distance from O to P and draw pc parallel to PC the triangles OPC and Opc will be similar, from which

$$\frac{OP}{Op} = \frac{OC}{Oc} = \text{a constant}$$

Since $\frac{OP}{Op}$ is constant for all positions of the mechanism $\frac{OC}{Oc}$ is also constant for all positions and the link pc remains parallel to the link PC throughout the entire stroke. Also since the point c is constrained to move in a straight line parallel to the stroke of the engine and the ratio of the crank OP to its connecting rod PC is the same as the ratio of the smaller crank Op to its connecting rod pc, the cross-head motion is reproduced correctly at every point of the stroke.

In Fig 13 let OP = PC = 18" and Op = pc = 3" then ab will

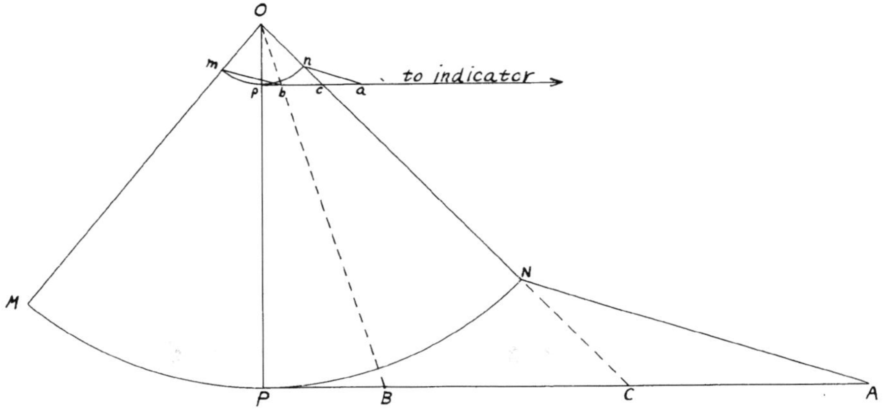

Fig. 13

equql 4" which is the length of the card. Fig. 14 is an assembly view of the rig showing the principle dimentions. Details of each of the separate pieces are shown in Fig. 15, (g) being the upper guide connection which is not shown in the assembly view. The pieces m m in Fig. 14 are guides for the rod r and are now in place on the engine making it unnecessary to detail them.

Bill of Material.

1 soft steel bar, 1" x ⅛" x 3' - 6"

1 soft steel bar, 1½" x ⅜" x 1' - 7"

1 soft steel bar, 1¼" x ⅜" x 2' - 0"

1 soft steel plate, 5" x ½" x 9¾"

1 hard steel rod, 1⅝" x 1¾"

1 hard steel rod, 1⅛" x 1½"

1 hard steel rod, ⅝" x 10"

4 brass washers, outside diameter, 1"
inside diameter, ½"
thickness 1/32"

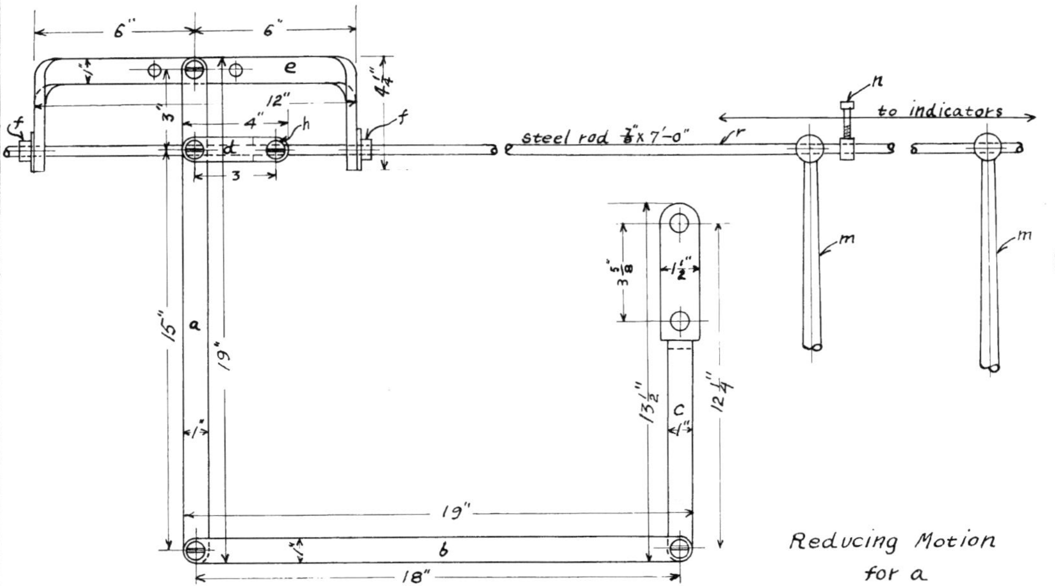

Fig. 14.

Reducing Motion
for a
12" x 24" Corliss Engine.

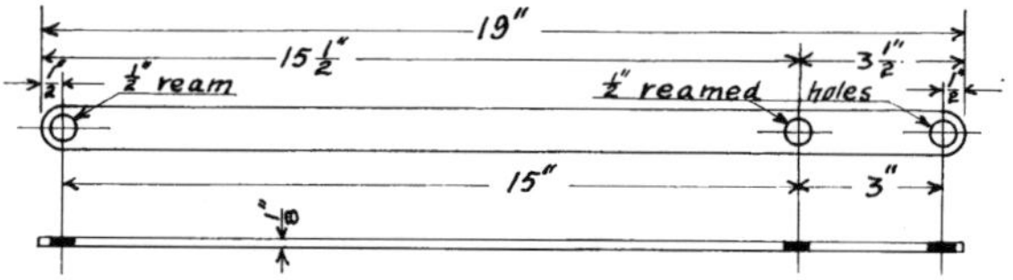

(a) Soft Steel, Make 1.

(b) Ditto (a) omit center hole.

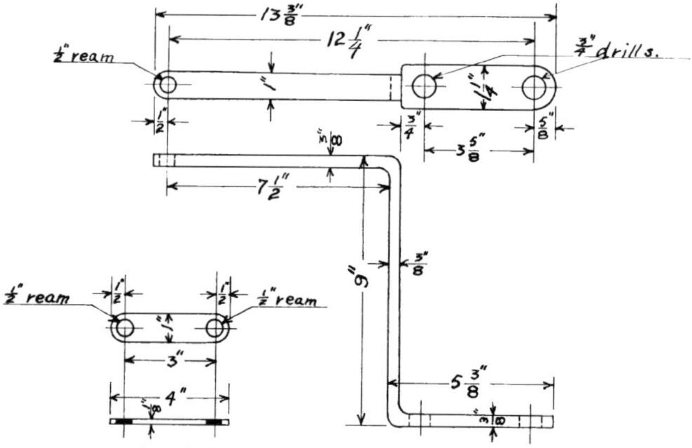

(d) Soft Steel, Make 1. (c) Soft Steel, Make 1.

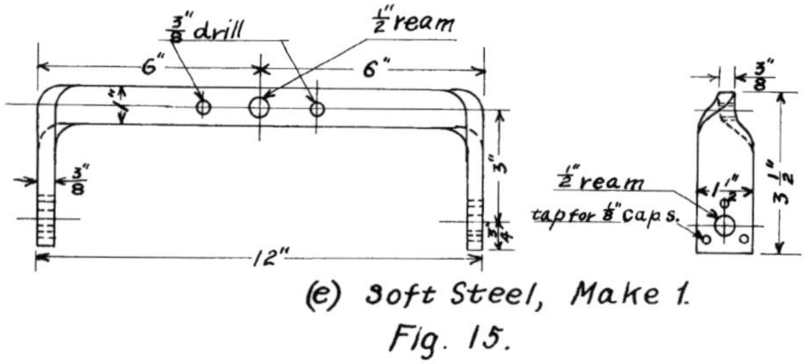

(e) Soft Steel, Make 1.

Fig. 15.

Continued on page 37.

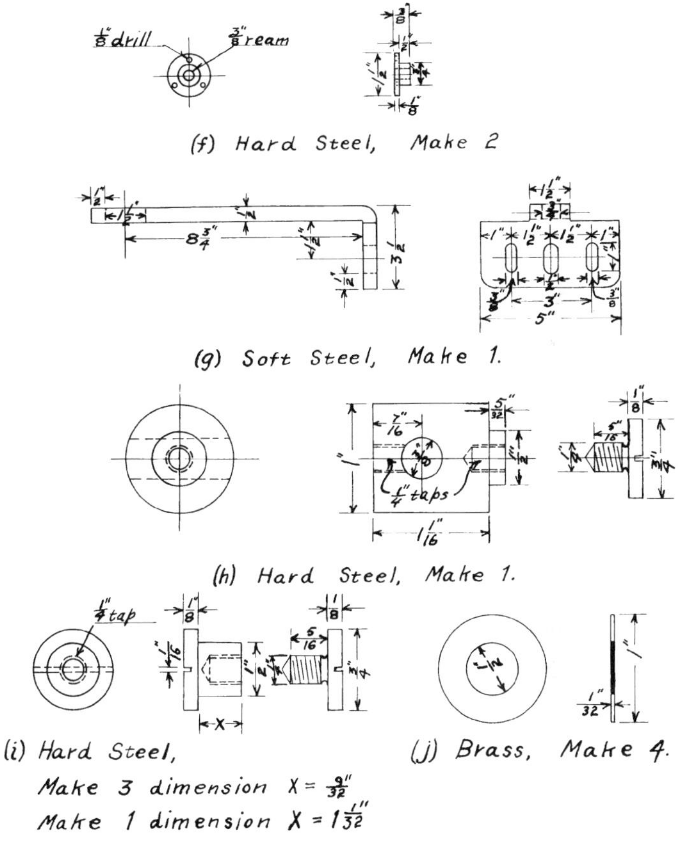

Fig. 15.

Details of Reducing Motion for a 12" X 24" Corliss Engine.

VII THE DESIGN OF A HELICAL REDUCING MOTION

Another form of reducing motion which may be used on a Corliss engine is that shown in Fig. 16. The engine for which this rig was designed is a 8" x 18" Murray-Corliss. An arm (b) is attached to the cross-head of the engine and slides upon the helical surface (a) causing the rod (r) to rotate through a small angle. Fastened to the rod (r) are two sectors (e) from which cords lead to the indicators. The helical surface is kept in contact with the arm (b) by means of a coil spring attached to some point on the engine frame. The rotation of the helical surface about its axis is proportional to the stroke of the engine for all positions of the cross-head and therefore the reduced motion is exact.

If we assume that the helix rotates through an angle of 60° or 1/6 of a revolution the pitch of the helix must be 6 times the length of the stroke;

Pitch of helix = 6 x 18" = 108" or 9'

Taking a sector $3\frac{3}{4}$" in diameter the length of the card is,

$$\frac{3.75 \times 60}{57.29} = 3.93''$$

Fig. 16 shows an assembly view of the rig and Fig. 17 is a detail drawing of the several parts.

Bill of Material.

1 cast iron helix, 22" long, $\frac{3}{4}$" thick, pitch 9'- 0"

1 steel rod, $\frac{3}{8}$" x 3'- 7"

2 cast iron sectors, 90° arcs, 1" thick, 4" radius.

1 soft steel bar, $1\frac{1}{2}$" x $\frac{1}{2}$" x 5'- 3".

1 coil spring, 1" x 1'- 0".

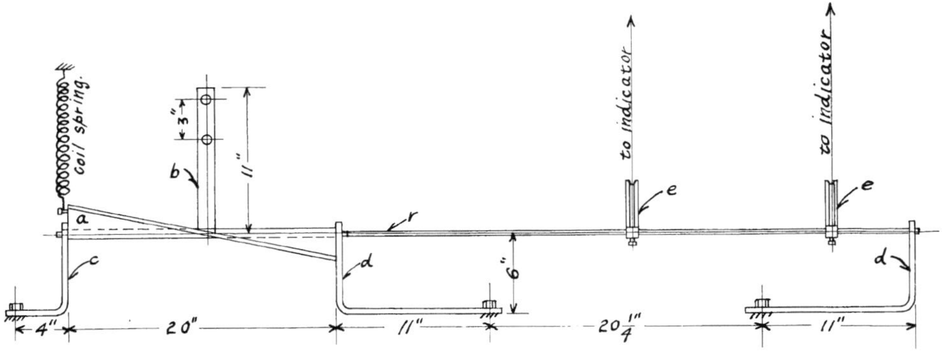

Fig. 16.

Helical Reducing Motion
for an 8"X18" Corliss Engine

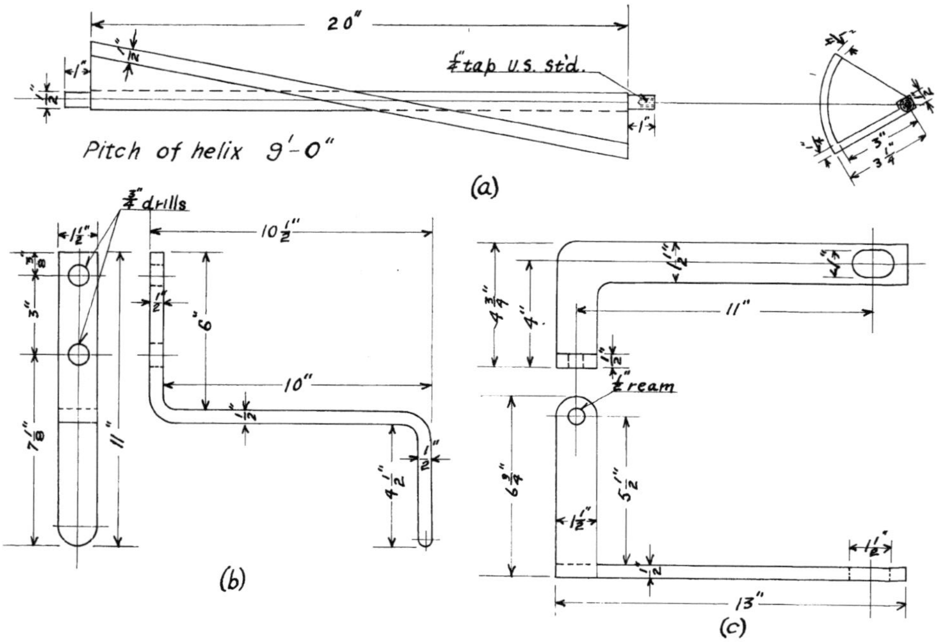

Fig. 17

Continued on page 41.

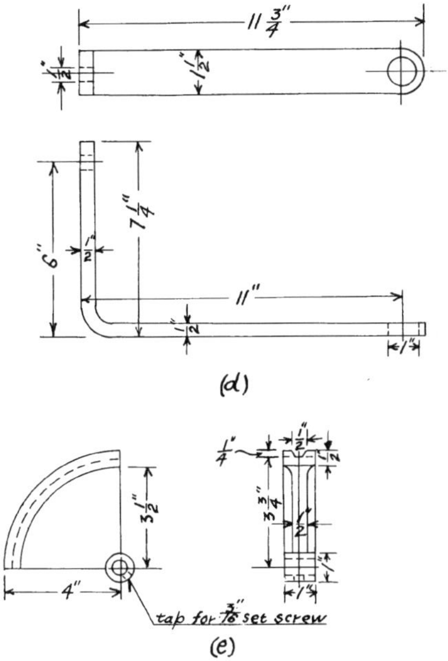

NOTE:

(a) Cast Iron Make 1.
(b) Soft Steel Make 1.
(C) Soft Steel Make 1.
(d) Soft Steel Make 2
(e) Cast Iron Make 2.

Fig. 17.

Details of Helical Reducing Motion.

Spotted something that isn't quite right?

Let us know and we'll fix it for future reprints.
And as a thank you for your diligence, we'll
send you a free Sothis book of your choice.

publisher@sothis.press

This page intentionally left blank

Except of course for that line above. And this one.
And whilst we have your attention, perhaps you'd
like to follow us on Instagram @sothis.press. Go on.

Made in United States
Orlando, FL
05 November 2026

90231076R10030